TRICERATOPS

by Laura K. Murray

Consultant: Mathew J. Wedel, PhD
Western University of Health Sciences
Pomona, California

PEBBLE
a capstone imprint

Published by Pebble, an imprint of Capstone
1710 Roe Crest Drive, North Mankato, Minnesota 56003
capstonepub.com

Library of Congress Cataloging-in-Publication Data
Names: Murray, Laura K., 1989- author.
Title: Triceratops / by Laura K. Murray.
Description: North Mankato, Minnesota : Pebble, an imprint of Capstone, [2025] | Series: Pebble explore | Includes bibliographical references and index. | Audience: Ages 5-8 | Audience: Grades 2-3 | Summary: "Readers want to know all about dinosaurs! They can get all the facts on the giant plant-eating dinosaur Triceratops. Engaging text and images make this book a great choice for information to use in a report or just to read for fun"— Provided by publisher.
Identifiers: LCCN 2024021174 (print) | LCCN 2024021175 (ebook) | ISBN 9780756588120 (hardcover) | ISBN 9780756589301 (paperback) | ISBN 9780756589172 (pdf) | ISBN 9780756589325 (kindle edition) | ISBN 9780756589318 (epub)
Subjects: LCSH: Triceratops—Juvenile literature. | Dinosaurs—Juvenile literature.
Classification: LCC QE862.O65 M896 2025 (print) | LCC QE862.O65 (ebook) | DDC 567.915/8—dc23/eng/20240614
LC record available at https://lccn.loc.gov/2024021174
LC ebook record available at https://lccn.loc.gov/2024021175

Editorial Credits
Editor: Erika L. Shores; Designer: Dina Her; Media Researcher: Jo Miller; Production Specialist: Tori Abraham

Image Credits
Alamy: Jim West, 26, Stocktrek Images, Inc., 10, Susan E. Degginger, 27, ZUMA Press Inc, 28; Capstone: Jon Hughes, cover, 1, 5, 6, 8, 11, 14, 17, 21, 23, 25; Getty Images: MARK GARLICK/SCIENCE PHOTO LIBRARY, 24; Science Source: JA CHIRINOS, 20; Shutterstock: Dotted Yeti, 18, Elenarts, 19, Kues, background (throughout), Ton Bangkeaw, 13, Zack Frank, 9

Printed and bound in the USA. 6618

Table of Contents

Words in **bold** are in the glossary.

Three-Horned Dinosaur

What dinosaur is known for its three horns? Triceratops! Triceratops was a big, slow-moving dinosaur. It had one of the biggest skulls of any land animal. Its name means "three-horned face."

Triceratops lived in the Late Cretaceous Period. That was about 68 million to 65 million years ago. Triceratops was one of the last dinosaurs.

Where in the World

Triceratops lived in North America. It roamed the western United States and Canada. In the U.S., it lived in Colorado, Montana, South Dakota, North Dakota, and Wyoming. In Canada, it lived in Alberta and Saskatchewan.

North America looked very different then. The **continent** was split into two separate lands. There was a sea between them.

Triceratops lived in places with lots of plants to eat. It lived in low, flat areas. It made its home near rivers, creeks, streams, or wetlands. Dry forests were full of plants. Today, these areas are **prairies** and **badlands**.

Many Triceratops **fossils** have been found in Montana. An area stretching across northeast Montana is famous for having lots of fossils. The land was formed by rivers, streams, and swamps.

Triceratops Bodies

Triceratops had a big, heavy body. It weighed 12,000 to 20,000 pounds (5,440 to 9,072 kilograms). It grew to 30 feet (9 meters) long. That is as long as a school bus!

Did You Know?

Triceratops was about 10 feet (3 m) tall. An adult man is about 6 feet (1.8 m) tall.

Triceratops stood on four legs. Its front
legs had three hooves each. Its back legs
were longer. They had four hooves each.
Triceratops had a pointed tail. Its skin
was made of large **scales**.

Triceratops had a huge skull. Its head took up almost one-third of its entire length! It had a sharp, horned beak.

Triceratops is best known for its three horns. Two long horns grew from its brow. A shorter horn grew near its nose. The horns may have been used for fights. Or they may have helped show off for **mates**.

Did You Know?

Scientists put Triceratops into two groups. One kind had a small nose horn. It had a long beak. The other kind had a long nose horn with a shorter beak.

frill

Triceratops had a huge, bony plate behind its head. It is called a frill. The frill was like a shield. It went over the dinosaur's neck. The frill had 19 to 26 small spikes on it.

The frill likely had many uses. It may have been used to fight **predators**. Triceratops may have fought each other too. The frill may have helped males show off to females.

What Triceratops Ate

Triceratops was a plant-eating dinosaur. It ate leaves and fruits. It ate roots, seeds, and twigs too.

Triceratops ate plants that grew close to the ground. It also used its big, strong body to help get food. Its horns helped reach taller plants.

Strong jaws and teeth were important for eating. The dinosaur's sharp beak could snip off tough plants. Its teeth were like scissors. They sliced up the plants.

Triceratops had up to 800 teeth at
a time. The teeth were long and sharp.
They were stacked on top of each other.
Older teeth got worn down. They fell out.
New ones grew in their place.

Life of Triceratops

A baby Triceratops was born from an egg. The baby may have been the size of a cat. Young Triceratops grew small, straight horns. The horns got bigger as the dinosaurs grew. The horns became curved.

Young Triceratops may have lived
with others. They learned to find
plants to survive.

Scientists look for clues to learn how dinosaurs lived. Some dinosaurs lived in herds. Their bones have been found near each other. Triceratops may have been different. Most Triceratops fossils have been found alone. A few young Triceratops fossils have been found together.

Triceratops did not have many predators. It was too big for most animals to attack. It could fight back with its big horns.

Tyrannosaurus rex ate meat from other dinosaurs. T. rex may have hunted young or weak Triceratops. Triceratops bones with T. rex bite marks have been found. But scientists do not know if T. rex and Triceratops fought often.

Discovering Triceratops

Scientists learn about dinosaurs from fossils. Fossils can be bones, teeth, and footprints. They can be shells, eggs, and more. People have found many fossils of Triceratops.

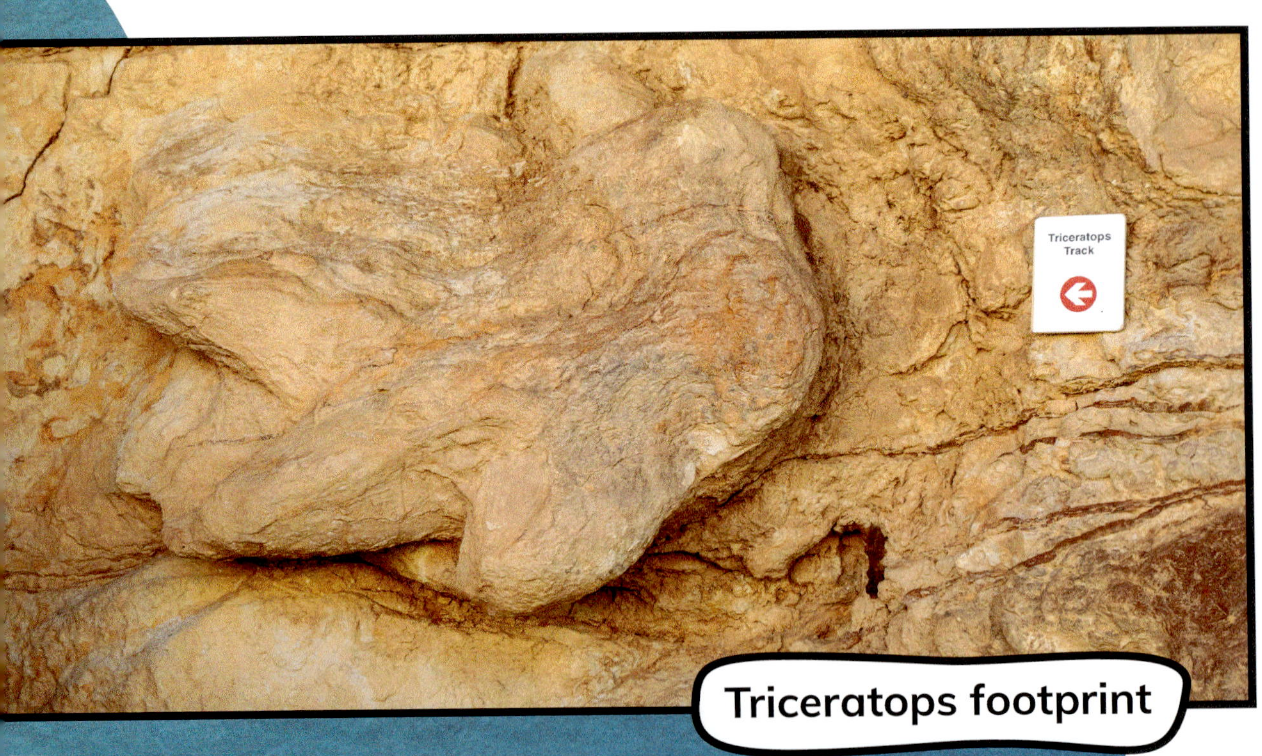

Triceratops footprint

The first Triceratops fossil was found in 1887. It was a skull with horns. George Cannon found the fossil near Denver, Colorado. He sent it to scientist O. C. Marsh. At first, Marsh thought the fossil was a bison. But soon he found out it was a dinosaur! Marsh gave the Triceratops its name in 1889.

The largest Triceratops fossil was found in South Dakota. The skeleton is called Big John. It weighs more than 1,500 pounds (680 kg). Another famous Triceratops is called Horridus. Its skeleton was found in Montana. It has 266 bones.

Today, people still study Triceratops. They look for clues about how Triceratops lived. There is a lot left to learn about these amazing three-horned dinosaurs!

Fast Facts

Name: Triceratops (meaning "three-horned face")

Lived: Late Cretaceous Period (about 68 to 65 million years ago)

Range: western United States (Colorado, Montana, South Dakota, North Dakota, Wyoming) and Canada (Alberta, Saskatchewan)

Habitat: low floodplains

Food: plants and plant material (leaves, fruit, roots, seeds, twigs)

Threats: Tyrannosaurus rex

Discovered: 1887, Colorado

Glossary

badlands (BAD-lands)—areas of dry land that have been worn away and have few plants

continent (KAHN-tuh-nuhnt)—one of Earth's seven large land masses

fossil (FAH-suhl)—the remains or traces of a living thing from many years ago

mate (MATE)—a partner that joins with another to produce young

prairie (PRAIR-ee)—a big area of flat land covered with grass

predator (PRED-uh-tur)—an animal that hunts other animals for food

scale (SKALE)—one of the small flat stiff plates that form an outer covering on the body of reptiles and some other animals

Read More

Nelson, Louise. *Creatures of the Cretaceous.*
Minneapolis: Bearport Publishing, 2023.

Throp, Claire. *Read All About Dinosaurs.*
North Mankato, MN: Capstone, 2022.

Vonder Brink, Tracy. *The Triceratops.* New York:
Crabtree Publishing, 2024.

Internet Sites

American Museum of Natural History: Dinosaur Facts
amnh.org/dinosaurs/dinosaur-facts

National Geographic Kids: Triceratops
kids.nationalgeographic.com/animals/prehistoric
/facts/triceratops

Natural History Museum: Triceratops
nhm.ac.uk/discover/dino-directory/triceratops.html

Index

About the Author

Laura K. Murray is the Minnesota-based author of more than 100 books for young readers. She loves learning from fellow readers and helping others find their reading superpowers! Visit LauraKMurray.com.